# GUIDE TO BECOMING A VETERINARIAN

I0446568

## Beginners Step-By-Step Guide To Navigating The Veterinarian's Path - A Vet's Journey

**JASE ROBBIN**

# Table of Contents

# Introductory

A veterinarian, abbreviated "vet," is an animal disease and injury specialist who specializes in the prevention, treatment, and diagnosis of such conditions.

Veterinarians possess the necessary expertise to administer medical treatment to an extensive array of fauna and plant life, in addition to domestic animals (including birds, horses, and canines), livestock (including pigs, cows, and horses), and exotic creatures.

A veterinarian may be entrusted with the following duties: conducting physical examinations, diagnostic tests, vaccination administration, medication prescription, surgical procedures, and the provision of nutritional and preventive care recommendations. Veterinarians fulfill an essential function in ensuring the overall health and welfare of animals, in addition to attending to public health issues pertaining to zoonotic diseases, which are those that are capable of transmission from animals to humans.

While some veterinarians treat companion animals and pets in private practice, others may specialize in pathology, public health, research, large animal medicine, or wildlife conservation.

Veterinary professionals may also be employed in academia, government agencies, or the private sector, where they contribute to diverse facets of animal welfare and health.

# CHAPTER ONE
## Prerequisites For Undergraduate Students

Elective veterinary schools and the country in which one intends to pursue veterinary education may have distinct undergraduate prerequisites. Notwithstanding this, numerous veterinary colleges generally enforce certain prerequisites.

It is vital to verify the particular admission requirements of the veterinary institutions that pique your interest, as these are merely general recommendations.

**Common undergraduate prerequisites for veterinary school include the following:**

1. **Bachelor's Degree:** A bachelor's degree is a prerequisite for admission to the majority of veterinary institutions. The degree may be awarded in any discipline; however, coursework in the sciences is typically emphasized.

2. **Prerequisite Courses:** In general, veterinary institutions mandate that prospective students have fulfilled specific prerequisite courses. Typical subjects covered in these courses comprise biology, chemistry, physics, and mathematics. Additionally, certain academic institutions might impose

academic prerequisites in fields including biochemistry, genetics, and statistics.

**3. Animal Experience:** Practical experience interacting with animals is highly regarded by numerous veterinary schools. Volunteering or working at zoos, veterinary clinics, animal shelters, or ranches are all viable options. This experience demonstrates your dedication to the animal care industry and your knowledge of the field.

**4. Recommendation Letters:** Letters of recommendation are frequently requested by veterinary institutions as an integral component of the application procedure. Professors, veterinarians, or other individuals who can attest to your scholastic prowess, work ethic, and suitability for a career in veterinary medicine typically compose these letters.

**5. Benchmark Assessments:** The Graduate Record Examination (GRE) is frequently mandated for veterinary school admission in certain nations, including the United States. Additionally, some institutions may mandate the Biology GRE subject test.

**6. Inclusion of Interviews:** Certain veterinary colleges may mandate an interview as a component of the admissions procedure. The admissions committee will have the opportunity to evaluate your professionalism, communication abilities, and drive to pursue a career in veterinary medicine.

In many cases, veterinary institutions impose minimum GPA criteria that must be met in order to be taken into account. This consists of the cumulative GPA as well as the GPA in each prerequisite course.

As these requirements can vary, it is vital to investigate the specific requirements of the veterinary colleges to which you are considering applying. In addition, maintaining a solid academic standing and acquiring experience in a veterinary-related field will increase your competitiveness as an applicant.

## Extracurricular Engagements

The inclusion of extracurricular activities in one's application to veterinary school can be of critical importance. Admissions committees seek candidates with a comprehensive skill set, including strong academic performance, dedication to the discipline of veterinary medicine, and effective interpersonal and leadership abilities. The following extracurricular activities can enhance your application to veterinary school:

1. **Practical Education:** Acquire practical experience in the field of animal care. Engaging in volunteer work or employment at veterinary clinics, animal shelters, farms, zoos, or

wildlife rehabilitation centers are all viable options. Gaining a wider range of experiences with various species is highly advantageous.

**2. Research:** Participate in research endeavors pertaining to animal science, biology, or veterinary medicine whenever feasible. A research background that showcases one's aptitude for analysis and resolution of challenges can be especially beneficial for individuals aspiring to enter the field of veterinary research.

**3. Internships or Externships:** To acquire practical experience and familiarity with the daily obligations of a veterinarian, pursue internships or externships in veterinary environments.

This may also afford you opportunities for mentorship.

**4. Assume Leadership Positions:** Engage in leadership activities within student organizations, societies, or community organizations. Proficient leadership experience serves as evidence of an individual's propensity for initiative, collaborative aptitude, and accountability.

**5. Clubs and Organizations:** Obtain membership in pre-veterinary clubs, animal-related clubs, or other pertinent organizations within your campus community. These organizations may provide opportunities for networking, guest speakers, and events that increase your exposure to the field.

**6. Community Service:** Participate in community service initiatives, with a particular emphasis on those that pertain to public health or animal welfare. The exhibition of a dedication to community service may garner a favorable evaluation from admissions committees.

**7. Enhancing Communication Abilities:** Cultivate your communication proficiencies by engaging in endeavors such as elocution, composition, or participation in debate collectives. Frequently, veterinarians must effectively communicate with the general public, colleagues, and clients.

**8. Engagement in Conferences and Workshops:** Attend seminars, workshops, and conferences that pertain to the field of veterinary medicine. This serves as an indication of your commitment to remaining updated on the latest advancements in the discipline and facilitates connections with others.

**9. Teaching or tuition:** Venture into the realm of tuition or instructing others if you possess a profound comprehension of specific subjects. This exemplifies your aptitude for effectively communicating information, a critical competency for an aspiring veterinarian.

**10. Engagement in Athletics or Creative Professions:** Active participation in athletics or the arts can serve as a platform to exhibit commendable attributes including diligence, collaboration, and ingenuity. These attributes hold significance in a multitude of veterinary medicine domains, encompassing client communication and surgical procedures.

Keep in mind that quality frequently takes precedence over quantity. Instead of attempting to engage in a multitude of activities, concentrate on a select few that authentically captivate your attention and afford you the opportunity to effect significant change. Emphasize your achievements and acquired proficiencies throughout your application and interviews. Veterinary institutions seek individuals who are multifaceted, have a deep affection for animals, and demonstrate a steadfast dedication to the veterinary field.

# CHAPTER TWO
## Tips For A Strong Application

Creating a strong application for veterinary school involves showcasing your academic achievements, relevant experiences, and personal qualities that make you a well-rounded candidate. Here are some tips to help you build a compelling veterinary school application:

1. **Maintain a Strong Academic Record:**

- Strive for a high GPA, especially in the prerequisite science courses.

- Demonstrate your ability to handle challenging coursework in the sciences.

2. **Gain Animal Experience:**

- Accumulate hands-on experience with animals through volunteering, internships, or work in veterinary clinics, shelters, farms, or zoos.
- Showcase the diversity of your experiences with different species.

3. **Research and Specialization:**

- Engage in research projects related to animal science or veterinary medicine, if possible.

- Clearly articulate any specific areas of interest or specialization within veterinary medicine.

4. **Demonstrate Leadership and Initiative:**

- Take on leadership roles in campus organizations, clubs, or community groups.
- Highlight instances where you demonstrated initiative and made a positive impact.

5. **Strong Letters of Recommendation:**

- Obtain letters of recommendation from professors, veterinarians, or professionals who can speak to your academic abilities, work ethic, and suitability for a career in veterinary medicine.

- Ensure your recommenders know you well and can provide specific examples of your strengths.

6. **Effective Communication Skills:**

- Develop and highlight your communication skills through public speaking, writing, or participation in activities that enhance your ability to convey information effectively.

7. **Show Community Engagement:**

- Participate in community service, especially activities related to animal welfare or public health.
- Demonstrate a commitment to serving and improving your community.

8. **Prepare for Standardized Tests:**

- Take required standardized tests, such as the GRE, and prepare adequately. Some schools may also require subject tests, so be sure to check specific requirements.

9. **Networking and Professional Development:**

- Attend conferences, workshops, and seminars related to veterinary medicine to stay informed about current developments and to network with professionals in the field.

10. **Highlight Unique Experiences:**

- Showcase any unique or impactful experiences you've had that set you apart from other applicants.
- Emphasize personal qualities, experiences, or challenges that have contributed to your growth

and preparedness for a career in veterinary medicine.

11. **Craft a Compelling Personal Statement:**

- Write a thoughtful and compelling personal statement that explains your motivation for pursuing veterinary medicine, your experiences, and your goals.
- Tailor your statement to the specific qualities and values of the veterinary school you're applying to.

12. **Practice for Interviews:**

- If interviews are part of the application process, practice answering common interview questions and be prepared to discuss your experiences, motivations, and goals in a clear and articulate manner.

Remember to thoroughly research the specific requirements of each veterinary school you are applying to, as they may have unique preferences and criteria. Tailor your application to align with the values and expectations of each school, and demonstrate how you will contribute to the veterinary community.

# Navigating Veterinary School

Navigating veterinary school can be both challenging and rewarding. Here are some tips to help you successfully navigate your veterinary education:

**Academic Success:**

1. **Stay Organized:**

- Use planners, calendars, or apps to keep track of assignments, exams, and deadlines.
- Develop effective time management skills.

2. **Active Learning:**

- Actively participate in class discussions, labs, and practical sessions.
- Form study groups to reinforce your understanding of complex topics.

3. **Effective Study Techniques:**

- Experiment with different study techniques to find what works best for you (flashcards, diagrams, practice questions, etc.).
- Develop a regular study routine to stay on top of coursework.

4. **Seek Help When Needed:**

- Don't hesitate to reach out to professors, tutors, or classmates if you're struggling with a particular concept.
- Utilize available resources such as study groups, office hours, and tutoring services.

## Clinical and Practical Experience:

1. **Maximize Clinical Rotations:**

- Actively participate in clinical rotations to gain hands-on experience and exposure to different aspects of veterinary medicine.

- Seek out diverse cases to broaden your skills.

2. **Build Relationships with Mentors:**

- Establish relationships with veterinary professionals who can provide guidance and mentorship.
- Ask questions and express your interests to faculty members and practicing veterinarians.

3. **Practice Professionalism:**

- Demonstrate professionalism in your interactions with faculty, staff, and peers.

- Develop strong communication skills, both written and verbal.

**Well-Being and Balance:**

1. **Maintain Physical Health:**

- Prioritize regular exercise, a balanced diet, and sufficient sleep to support your overall well-being.
- Manage stress through healthy coping mechanisms.

2. **Emotional Support:**

- Cultivate a support network of friends, family, and fellow students.

- Seek counseling services if needed, as veterinary school can be mentally challenging.

3. **Work-Life Balance:**

- Strive for a balance between academic commitments and personal life.
- Set realistic goals and boundaries to avoid burnout.

## Professional Development:

1. **Networking:**

- Attend conferences, workshops, and networking events to connect with professionals in the field.

- Join professional organizations related to veterinary medicine.

2. **Stay Informed:**

- Keep up with industry developments, research, and advancements in veterinary medicine.
- Subscribe to relevant publications and journals.

3. **Prepare for Licensure:**

- Understand the licensure requirements in the region where you plan to practice.
- Prepare for licensing exams as required.

**Career Planning:**

1. **Explore Specializations:**

- Take advantage of opportunities to explore different specializations within veterinary medicine.
- Seek guidance from mentors and professionals in your areas of interest.

2. **Internships and Externships:**

- Pursue internships or externships to gain additional hands-on experience and make informed career decisions.
- Network with professionals in your desired field.

3. **Job Search and Preparation:**

- Start preparing for your job search well before graduation.
- Attend career fairs, workshops, and utilize career services at your institution.

Remember, veterinary school is a significant investment of time and effort, so make the most of your education by actively engaging in your studies, seeking valuable experiences, and maintaining a healthy balance between academic and personal life.

# CHAPTER THREE
## Veterinary Specializations

Veterinary medicine offers a wide range of specializations, allowing veterinarians to focus on specific areas of expertise. Here are some common veterinary specializations:

1. **Internal Medicine:**

- Veterinarians specializing in internal medicine focus on the diagnosis and treatment of diseases affecting internal organs, such as the heart, liver, kidneys, and gastrointestinal tract.

2. **Surgery:**

- Surgical specialists perform various types of surgeries, including soft tissue and orthopedic procedures. They may work on complex cases, such as tumor removals or joint surgeries.

3. **Dentistry:**

- Veterinary dentists specialize in oral health, treating conditions related to the teeth, gums, and oral cavity. They may perform dental cleanings, extractions, and other dental procedures.

4. **Emergency and Critical Care:**

- Veterinarians in this specialization focus on treating animals with critical or life-threatening conditions. They often work in emergency clinics or hospitals and are trained to handle urgent medical situations.

5. **Cardiology:**

- Cardiologists specialize in the diagnosis and treatment of heart-related issues in animals. They may work with conditions such as heart murmurs, congestive heart failure, or congenital heart defects.

6. **Dermatology:**

- Veterinary dermatologists focus on skin, ear, and coat conditions in animals. They diagnose and treat issues like allergies, infections, and autoimmune skin diseases.

7. **Ophthalmology:**

- Ophthalmologists specialize in eye care for animals. They diagnose and treat conditions affecting the eyes, including cataracts, glaucoma, and corneal disorders.

## 8. **Radiology:**

- Veterinary radiologists use imaging techniques such as X-rays, ultrasound, and MRIs to diagnose and treat various conditions. They may work closely with other specialists to interpret diagnostic images.

## 9. **Neurology:**

- Neurologists focus on disorders of the nervous system in animals. This can include conditions such as seizures, spinal cord injuries, and neurological diseases.

10. **Anesthesiology:**

- Anesthesiologists specialize in administering anesthesia and managing pain during surgical and medical procedures. They play a crucial role in ensuring the safety and well-being of animals undergoing various treatments.

11. **Pathology:**

- Veterinary pathologists examine tissues, organs, and bodily fluids to diagnose diseases. They often work in laboratories and contribute to research and disease surveillance.

12. **Behavior:**

- Veterinary behaviorists specialize in the study and treatment of behavior problems in animals. They may address issues such as aggression, anxiety, or compulsive behaviors.

13. **Zoological Medicine:**

- Veterinarians in this specialization work with exotic and zoo animals. They may be involved in the care, conservation, and research of wildlife and animals in captivity.

14. **Equine Medicine:**

- Equine veterinarians focus on the health and well-being of horses. They may work with performance horses, racehorses, or horses in various disciplines, addressing issues ranging from lameness to reproductive health.

These are just a few examples of the many veterinary specializations available. Aspiring veterinarians interested in a particular field can pursue additional training, internships, or residencies to gain expertise in their chosen area.

The field of veterinary medicine continues to evolve, offering opportunities for veterinarians to contribute to various aspects of animal health and welfare.

## Clinical Skills And Techniques

Clinical skills and techniques are fundamental for veterinarians to diagnose, treat, and care for animals effectively.

These skills encompass a wide range of abilities, including physical examination, diagnostic procedures, surgical techniques, and communication with clients.

Here are some essential clinical skills and techniques for veterinarians:

**Physical Examination:**

1. **Palpation:**

- Assessing the texture, size, and consistency of tissues through touch. This is crucial for detecting abnormalities in organs and structures.

2. **Auscultation:**

- Listening to internal sounds using a stethoscope. This is commonly done to evaluate heart and lung function.

3. **Percussion:**

- Tapping the body surface to assess underlying structures. This can help identify abnormalities in the abdomen or chest.

4. **Observation:**

- Careful visual examination to detect signs of illness or injury, such as changes in posture, gait, or behavior.

**Diagnostic Procedures:**

5. **Radiography (X-rays):**

- Capturing images of internal structures to diagnose

conditions like fractures, tumors, or organ abnormalities.

6. **Ultrasound:**

- Using sound waves to create real-time images of internal organs. This is valuable for evaluating soft tissues and detecting abnormalities.

7. **Endoscopy:**

- Inserting a flexible tube with a camera into body cavities to visualize and diagnose issues without invasive surgery.

8. **Blood and Urine Analysis:**

- Conducting laboratory tests on blood and urine samples to assess organ function, detect infections, and identify metabolic disorders.

**Surgical Techniques:**

9. **General Surgery:**

- Performing routine surgical procedures such as spaying, neutering, and tumor removals.

10. **Orthopedic Surgery:**

- Addressing musculoskeletal issues, including fractures, joint problems, and ligament injuries.

11. **Soft Tissue Surgery:**

- Addressing issues in organs and tissues, such as gastrointestinal surgeries, tumor removals, and reconstructive procedures.

12. **Dentistry:**

- Conducting dental procedures, including cleanings, extractions, and addressing oral health issues.

**Anesthesia and Pain Management:**

13. **Administering Anesthesia:**

- Safely inducing and maintaining anesthesia during surgical procedures or diagnostic imaging.

14. **Pain Assessment and Management:**

- Evaluating and addressing pain in animals through appropriate medications and techniques.

**Communication and Client Interaction:**

15. **Client Education:**

- Effectively communicating with pet owners to explain diagnoses, treatment plans, and preventive care.

16. **Compassionate Communication:**

- Developing empathy and sensitivity when delivering difficult news or discussing end-of-life decisions.

17. **Team Collaboration:**

- Working collaboratively with veterinary technicians, assistants, and other staff members to provide comprehensive care.

**Emergency and Critical Care:**

18. **Emergency Stabilization:**

- Responding to and stabilizing animals in critical conditions, including trauma, toxin exposure, or severe illness.

19. **CPR (Cardiopulmonary Resuscitation):**

- Administering life-saving measures in cases of cardiac or respiratory arrest.

**Continuous Learning and Adaptability:**

### 20. Keeping Up with Advances:

- Staying informed about new technologies, research, and treatments through continuing education.

### 21. Problem-Solving Skills:

- Approaching diagnostic challenges and complex cases with critical thinking and problem-solving skills.

Developing proficiency in these clinical skills and techniques requires a combination of formal education, hands-on training, and practical experience. Veterinarians often continue to refine and expand their skills throughout their careers to provide the best possible care for their animal patients.

# CHAPTER FOUR
## Professional Development

Professional development is a continuous process that involves acquiring and honing the skills, knowledge, and behaviors necessary to excel in one's chosen field. For veterinarians, professional development is essential to stay current with advancements in veterinary medicine, maintain competence, and adapt to evolving industry standards. Here are key aspects of professional development for veterinarians:

### 1. Continuing Education:

- Attend workshops, conferences, and seminars to stay informed about the latest research, treatments, and technologies in veterinary medicine.

- Participate in webinars or online courses that offer convenient ways to enhance knowledge and skills.

2. **Specialization and Certification:**

- Consider pursuing specialized training or certification in a particular veterinary field of interest. Specializations can include surgery, dentistry, pathology, or internal medicine.
- Achieving board certification in a specialty area demonstrates

advanced expertise and dedication to a specific field.

3. **Networking:**

- Engage with colleagues, mentors, and professionals in the veterinary community to foster connections and stay updated on industry trends.
- Join professional organizations and attend local or national events to expand your network and exchange ideas with peers.

4. **Mentorship:**

- Seek out mentorship from experienced veterinarians who

can provide guidance, share insights, and offer career advice.

- Consider mentoring veterinary students or younger professionals to contribute to the development of the next generation.

5. **Research and Publications:**

- Contribute to research projects, case studies, or publications in veterinary journals to share knowledge with the broader community.
- Stay involved in research activities to contribute to advancements in veterinary science.

6. **Utilizing Technology:**

- Embrace technology tools and platforms that can enhance veterinary practice management, communication with clients, and diagnostic capabilities.
- Stay updated on software, apps, and other technological advancements relevant to the field.

7. **Practice Management and Leadership:**

- Develop skills in practice management, including financial management, client relations, and team leadership.

- Consider courses or workshops on leadership to enhance your ability to manage a veterinary practice effectively.

8. **Ethical and Professional Behavior:**

- Adhere to ethical standards and codes of conduct in veterinary medicine.
- Stay informed about legal and regulatory requirements affecting veterinary practice.

9. **Well-Being and Work-Life Balance:**

- Prioritize personal well-being to prevent burnout and maintain a healthy work-life balance.
- Seek support and resources to address stress, compassion fatigue, and mental health challenges.

10. **Public Outreach and Education:**

- Engage in community outreach and education initiatives to promote responsible pet ownership, public health, and awareness of veterinary issues.

- Participate in educational programs for schools or community groups to raise awareness about the veterinary profession.

11. **Professional Reflection:**

- Regularly reflect on your career goals, achievements, and areas for improvement.
- Set personal and professional development goals to guide your continuous improvement.

By actively participating in these aspects of professional development, veterinarians can ensure they remain skilled, knowledgeable, and equipped to provide high-quality care to their animal patients while advancing their careers.

## Career Paths In Veterinary Medicine

Veterinary medicine offers a diverse range of career paths beyond traditional clinical practice. Here are some common career paths within the field of veterinary medicine:

1. **Clinical Practice:**

- **General Practice:** Work in private clinics providing care to companion animals, such as dogs and cats.
- **Specialty Practice:** Focus on a specific area of veterinary medicine, such as surgery,

dentistry, dermatology, or internal medicine.

2. **Research and Academia:**

- **Veterinary Research:** Contribute to scientific advancements in veterinary medicine through research in academia, industry, or government institutions.

- **Teaching:** Educate future veterinarians as a faculty member at a veterinary school or university.

3. **Public Health:**

- **Epidemiology:** Investigate and control the spread of diseases, monitor health trends, and contribute to public health policy.
- **Food Safety Inspection:** Ensure the safety of the food supply by inspecting and regulating food production facilities.

4. **Government and Regulatory Agencies:**

- **Veterinary Inspector:** Work for government agencies to inspect and regulate animal health, welfare, and food safety.
- **Disease Control Specialist:** Monitor and control the spread

of animal diseases at local, national, or international levels.

5. **Wildlife and Conservation:**

- **Wildlife Veterinarian:** Work with wildlife populations, including captive and free-ranging animals, to promote conservation and address health issues.
- **Zoo Medicine:** Provide veterinary care to animals in zoos, aquariums, and wildlife parks.

6. **Equine Practice:**

- **Equine Veterinarian:** Specialize in the health and well-being of horses, including performance animals, racehorses, and companion horses.

7. **Industry and Pharmaceutical Companies:**

- **Product Development:** Contribute to the development of veterinary pharmaceuticals, vaccines, and other products.
- **Sales and Marketing:** Work in sales or marketing roles for veterinary products or services.

8. **Emergency and Critical Care:**

- **Emergency and Critical Care Veterinarian:** Provide urgent care and treatment for animals in emergency situations, often in specialized clinics or hospitals.

9. **Pathology:**

- **Clinical Pathologist:** Analyze laboratory samples, such as blood, tissues, and fluids, to diagnose diseases.

10. **Public Education and Advocacy:**

- **Animal Welfare Officer:** Work for animal welfare organizations to ensure the well-being of animals and advocate for humane treatment.

- **Public Education:** Educate the public about responsible pet ownership, animal welfare, and veterinary issues.

11. **Reproductive Medicine:**

- **Reproductive Specialist:** Focus on reproductive health, including breeding programs, artificial insemination, and fertility assessments in animals.

12. **Therapeutic Services:**

- **Physical Rehabilitation Therapist:** Provide rehabilitation services for animals recovering from surgeries, injuries, or chronic conditions.

13. **Holistic and Alternative Medicine:**

- **Holistic Veterinarian:** Offer alternative therapies such as acupuncture, chiropractic care, and herbal medicine alongside traditional veterinary practices.

14. **Global and International Health:**

- **International Veterinary Medicine:** Contribute to global health initiatives, disease control, and animal welfare on an international scale.

15. **Telemedicine and Telehealth:**

- **Telemedicine Consultant:** Provide veterinary consultations and advice remotely, leveraging technology to connect with clients.

These are just a few examples, and the field of veterinary medicine continues to evolve, offering new opportunities and specialties. Veterinarians can also pursue hybrid career paths or combine multiple areas of interest to create a unique career that aligns with their passion and expertise.

# CHAPTER FIVE
## Challenges And Rewards

Veterinary medicine is a rewarding yet challenging profession that involves caring for the health and well-being of animals. Here are some of the challenges and rewards associated with a career in veterinary medicine:

**Challenges:**

1. **Emotional Stress:**

- Dealing with illness, injury, and sometimes euthanasia of animals can be emotionally challenging for veterinarians.

- Communicating with pet owners during difficult situations

requires empathy and emotional resilience.

2. **Work-Life Balance:**

- Veterinarians may work long hours, including evenings, weekends, and holidays, especially in emergency and critical care settings.
- Balancing work demands with personal life can be challenging, leading to potential burnout.

3. **Financial Strain:**

- The cost of veterinary education can result in substantial student loan debt.

- Veterinarians working in certain fields, such as public health or nonprofit organizations, may face financial constraints compared to those in private practice.

4. **Physical Demands:**

- Performing physical tasks, such as lifting and restraining animals, can lead to occupational injuries.
- Working in challenging environments, including outdoor settings or with large animals, can be physically demanding.

5. **Client Communication:**

- Communicating effectively with pet owners, including conveying complex medical information and discussing treatment options, requires strong interpersonal skills.
- Addressing client expectations and financial constraints can be challenging.

6. **Ethical Dilemmas:**

- Veterinarians may encounter ethical dilemmas related to treatment decisions, client expectations, and quality of life considerations.

- Balancing the best interests of the animal with client preferences can be challenging.

7. **Constant Learning:**

- The field of veterinary medicine is continually evolving, requiring veterinarians to engage in lifelong learning to stay current with advancements and new technologies.
- Keeping up with a broad range of medical specialties can be demanding.

**Rewards:**

1. **Animal Bond:**

- Building strong connections with animals and contributing to their well-being is a deeply rewarding aspect of veterinary medicine.

- Helping animals recover from illness or injury and supporting their overall health brings a sense of fulfillment.

2. **Variety of Work:**

- Veterinarians have the opportunity to work with various species, from companion animals to exotic wildlife, providing a diverse and stimulating work environment.

- Different clinical cases and challenges contribute to a varied and interesting career.

3. **Impact on Public Health:**

- Veterinarians play a crucial role in safeguarding public health by monitoring and controlling infectious diseases that can be transmitted between animals and humans.

- Ensuring food safety and inspecting animal products contribute to public health and safety.

4. **Human-Animal Bond:**

- Facilitating and preserving the bond between animals and their owners is a rewarding aspect of veterinary practice.
- Helping families care for their pets and witnessing the positive impact animals have on people's lives is fulfilling.

5. **Problem-Solving:**

- Diagnosing and treating medical issues in animals involve critical thinking and problem-solving skills.
- Overcoming challenges and finding solutions can be intellectually satisfying.

6. **Professional Relationships:**

- Building professional relationships with colleagues, mentors, and fellow veterinarians provides a supportive network.
- Collaborating with a team dedicated to animal care fosters a sense of community within the veterinary profession.

7. **Continuous Learning:**

- The dynamic nature of veterinary medicine allows for continuous learning and professional development.
- Staying informed about new research, technologies, and

treatment options contributes to professional growth.

Despite the challenges, many veterinarians find the rewards of making a positive impact on the lives of animals and their owners to be personally and professionally fulfilling. The balance between challenges and rewards can vary depending on the individual's preferences, career path, and chosen veterinary specialization.

## Summary

Veterinarian work is both demanding and fulfilling, necessitating an exceptional combination of scientific acumen, clinical expertise, and empathy.

Veterinary professionals maintain the human-animal bond while contributing to public health and safeguarding the health and well-being of animals. Nevertheless, the vocation presents a series of obstacles, such as psychological strain, extended periods of labor, and the imperative for ongoing education.

Although there are obstacles to overcome, a veterinary vocation offers substantial benefits. Developing robust relationships with animals, positively influencing their well-being, and fostering the bond between companion animals and their guardians are extraordinarily gratifying facets of the vocation.

Veterinarians are afforded the opportunity to engage in problem-solving, interact with a diverse array of species, and make significant contributions to the field of veterinary science.

In addition to technical proficiency, veterinary medicine frequently demands effective interpersonal and communication abilities. Maintaining a work-life balance, addressing ethical dilemmas, and managing client expectations are all responsibilities that veterinarians must confront.

A commitment to lifelong learning is an essential component of the vocation, given the perpetual advancements in research, technologies, and therapeutic alternatives.

In essence, those desiring to embark on a profession in veterinary medicine ought to possess a fervent regard for animals, an unwavering commitment to their welfare, and the capacity to withstand obstacles.

Veterinarians are able to customize their professional trajectories in accordance with their distinct aptitudes and areas of interest, encompassing clinical practice, research, public health, and education.

Although the pursuit of veterinary medicine may involve significant challenges, those who are motivated by the desire to improve the health of both animals and humans and the positive influence it has on the lives of animals will find it to be a rewarding and purposeful vocation.

**THE END**